The Nonsense of Global Warming and Climate Change

Dirk van Leenen

The Nonsense of Global Warming and Climate Change
Copyright © 2023 by Dirk van Leenen

All rights reserved. No part of this publication may be reproduced, distributed, or transmitted in any form or by any means, including photocopying, recording, or other electronic or mechanical methods, without the prior written permission of the author, except in the case of brief quotations embodied in critical reviews and certain other non-commercial uses permitted by copyright law.

Tellwell Talent
www.tellwell.ca

ISBN
978-0-2288-9640-1 (Hardcover)
978-0-2288-9639-5 (Paperback)

Introduction

The more I was hearing the words: **"Global Warming and Climate Change",** I realized that there was more behind these words.

As a Horticulturist, I travelled The Earth and experienced many different weather situations. Storms, floods, hurricanes, earthquakes and many different temperatures. I became more aware and in awe, how The Earth corrected all these, sometimes catastrophic events in a relatively short period of time. Not only The Earth, but even the people adjusted themselves to the situation.

After studying a lot of Global Warming and Climate Change literature, I realized that all the rules and regulations created by the Government had a different purpose than curing the natural events, which had been happening on The Earth as long as we could possibly remember.

I discovered that the reason was power and money. To grab a hold on the citizen, I call it a Power Grab! From that point on in my thinking, I realized that it was all about the Truth which was hidden in their actions.

I intensified my studies and found many more reasons for the un-truths as well as many scientists who condemned the Government actions. More scientist don't believe in Global Warming and Climate Change as it is portrayed by the Media.

This is what motivated me to write this book because Global Warming and Climate Change is unbelievable **NONSENSE!**

The Author, Dr. Dirk van Leenen

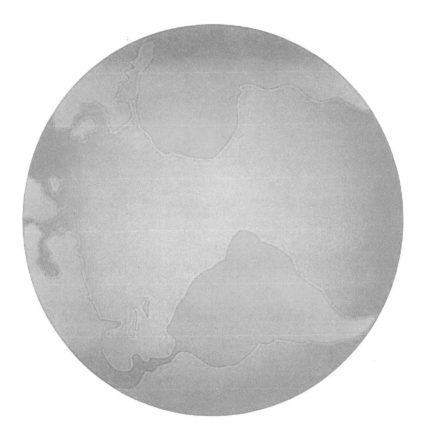

The Green Earth

Our complicated earth is unique in the Universe.

It is completely self-supporting, self-correcting, and self-improving. It is amazing that everyone who lives on this earth, takes The Earth as it is, for granted. As far as we know our earth is the only living planet in the unlimited Universe.

The only planet which is inhabited by intelligent beings. Of course, there are rumors and unconfirmed "evidence" that there are other planets with intelligent life, but nothing has ever been proven.

No man can change anything on this earth. Even though we think so. We cannot change Volcanos to erupt, earth quakes to occur, stop tsunamis, hurricanes, or tornadoes. The Earth corrects them all in due time. Even forest fires are a perfect example of corrections. Within 3 years the result of a fire is, that the land is more fertile and completely green again.

The earth supplies all our needs. Everything which we create, build, eat, use, enjoy, experience, is supplied by The Earth. It has been done for many thousands of years. Everything which The Earth supplies, returns to The Earth.

The present 7,753 billion population on this Earth is all alive and stays alive on it.

Dirk van Leenen

There is no other planet like this.

Louis Armstrong sang "What a wonderful World," Maybe he should have sung "What a wonderful Earth."

Astronaut William Anders remarked:" We're in awe about the beautiful Earth from space" and he captured a picture which became known as: "Earthrise."

Astronaut Jim Lovell said: "The vast loneliness up here of the Moon is awe inspiring, and it makes you realize just what you have back there on Earth. The Earth from here, is a grand oasis to the big vastness of space."

It is so very green!

Water Cycle on and above the Earth.

The water cycle describes how water evaporates from the Earth's surface, rises into the atmosphere, cools, condenses to form clouds, and falls again to the surface as precipitation. evaporation of water from the earth's surface. On land, water evaporates from the ground, mainly from soils, plants (i.e., transpiration), lakes, streams, rivers, and oceans. In fact, approximately 15% of the water entering the atmosphere is from evaporation from the Earth's land surfaces and evapotranspiration from plants.

Such evaporation cools the Earth's surface, cools the lower atmosphere, and provides water to the atmosphere to form clouds. So, in order to raise the global temperature men would have to change the temperature of the clouds and all the surface waters, and that is too big for our britches.

Fresh water on the Earth

Of all the water that exists on our planet, roughly 97 % is salt water and. less than 3% is fresh water.

Most of it is frozen in glaciers, ice caps, or is deep underground in aquifers.

Less than 3% of the Earth's water is fresh water which is easily accessible to us to meet our needs, and most of that water is replenished by precipitation, a vital component of the water cycle, affecting every living thing on earth.

About 75% of the energy (or heat) in the global atmosphere is transferred through

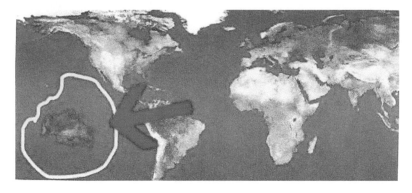

The Earth's Great Pacific Garbage Patch.

The earth operates on a cyclical basis.

Everything on the Earth is cyclical. That means, that everything that happens has happened for thousands of years and will happen again. And everything that has happened, occurred several times before in thousands of years. Knowing that, also entails, that the same happenings will occur again in the future.

Because only in the last two hundreds of years we have been able to begin to record all natural events, but we do not know much about what happened before that. For instance, fires have been burning since the beginning and they will keep happening over and over. Forest fires are ugly, but nature replaces them to brand-new (no pun intended) forests.

The earth uses all the products from a fire for the benefit of new growth. A fire eliminates insects that harmed the trees, e. g. Bark bugs, wood worms etc. The ashes from the fire turn into fertilizers, the burned wood becomes good mulch, the nitrogen and potash become the best fertilizer to regrow the trees and other vegetation.

The Earth is at work to correct any calamity by itself.

What does the earth contain?

To get an idea about the enormity of the earth and its contents, here are some statistics. The Earth has a Cubic Mileage of 260 billion cubic miles.

Total land mass: 58 million square miles, or 29 % of the total surface.

Total surface of water 71%, of which 97% is salt water and only 3% of fresh water.

Total Cubic miles of snow and ice 6 million cubic miles. Some scientist say that it would take 5000 years to melt.

Total surface of buildings and roads just 1%.

Gasses, on the earth are as follows: Oxygen (we breathe) 21%. Nitrogen78 %. Argon 0.9 % Other gasses 0.1%

CO_2 amounts to 0.4%, and it has been like that for hundreds of years, as far back as we have recorded it, but for thousands of years while it had not been recorded.

Dirk van Leenen

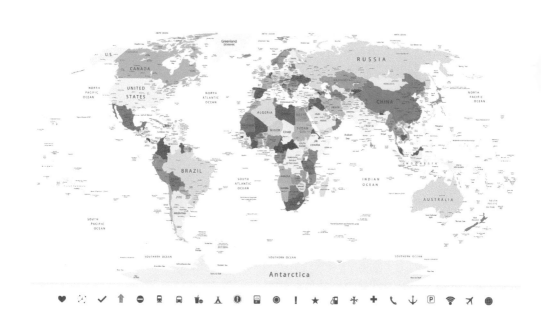

The layers around the Earth.

The earth is surrounded and protected by five mayor layers. They are like rings around The Earth. And they have names and certain distances from The Earth:

First layer is called **The Troposphere**, it is about 9 miles wide and contains the Oxygen we breath and the clouds that bring us precipitation. The air is at its densest at this layer.

Second Layer is the **Stratosphere**, this layer is about 22 miles thick. The air is much thinner, it contains: less Oxygen and Nitrogen and, contra the opinion of certain scientist hardly any CO2 can get there.

Here is where the controversy of global warming begins. In this layer we can find Ozone CO3, which protects humans from the sun's UV radiation. Since a conference was held in Vienna in 1985 on the depletion of Ozone not much has changed in the ozone concentration.

Scientist blamed certain man-made chemicals for the depletion of Ozone, but the question remains, how do those chemicals get to the stratosphere at 10 to 50 miles away from The Earth.

Where there is hardly any gravity towards the earth and those culprit fumes or gasses are heavier than the air we breathe. The situation was grossly overstated. Interestingly, the greatest depletion occurred in the Arctic and the North pole (where hardly any harmful chemicals are produced.) **Go figure!**

The third Layer is the **Mesosphere**, it is 22 miles thick about 50 to 85 miles away from The Earth.

Here the gasses are all mixed up. The air is too thin to breathe. The temperature is from -90 degrees, -130 degrees F. The sphere contains liquid iron and nickel.

The Mesosphere is our protection against meteors, which are braking up in the varied gasses. Most of the space debris, such as abandoned satellites are dissolved and that is why that debris never returns to our Earth.

It is a wonderful guard for the earth because meteors and meteorites could damage The Earth if they would be able to pass through the Mesosphere.

The fourth layer is the **Thermosphere**, it is 31 miles wide. It absorbs the sun's radiation which makes it very hot. Thermos, meaning Heat which can reach F 680 degrees even up to F4000 degrees.

Ultraviolet radiation coming from the sun causes **Photoionization** creating **ions** at only 50 miles above sea level.

Again, this layer is a protector for The Earth because there are not enough gasses to increases a transfer of the heat to The Earth.

Realize that the further from the Earth the Spheres are, the cubic volume drastically

The **Ionosphere** is part of the Thermosphere. Here the radiation causes atmospheric particles to become electrically charged which enables radio waves to be refracted and thus be received beyond the horizon. This enables the World to use it for telecommunication.

The fifth layer is the **Exosphere,** it is 6200 miles thick and thus the outermost region of the planet's atmosphere. It is also the first layer to come into contact and protect The Earth from meteors, asteroids and cosmic rays, the temperatures vary between F 0 and F 1700 degrees. This sphere is as big as the Earth itself, it contains Hydrogen, Helium in a very low density.

The exosphere is the transition zone between **The Earth and Outer Space.** Here is where all our satellites are.

(Picture of the Speres)# 3 again

The World

Pollution, Carbon Footprints, Greenhouse Gasses, etc. Most people do not realize that **The World** is a different entity than **The Earth,** for that reason I am explaining the difference in order to understand what we are dealing with when we talk about topics like Global Warming, Climate Change, New Green Deal,

The World is everything which is **on** The Earth, it is everything **men-made**, therefore The World is not self-correcting, self-healing of self-improving like The Earth is.

This entails that every road, every harbor, every airport, every building, every factory, every machine needs to be repaired by men, improved by men, maintained by men, reinvented by men just to mention a few. For all the activities, **we need everything from The Earth.**

Nothing can be made or it is coming from the earth.

The World is also the people on the Earth, and everything they eat, build, manufacture comes from The Earth. Whether we grow it, mine it, pump it, or dig it up, breed it, invent it, manufacture it, everything originates from the earth. All Chemicals, minerals, elements are used by **The World** and come from **The Earth**.

IT IS ALL HERE. BUT everything returns to the earth.

A famous saying is: **From Dust to Dust!** Later we will discuss Pollution.

We have developed many "isms." Here is a list of those that are used most frequently in the present time: Communism, Socialism, Capitalism, Fascism, Marxism Materialism, Liberalism, Economism, Conservatism. **These are all political isms.**

Some newly used **Isms:**
Fantasism, Greenism, Wokeism,
Climatism, Environism, Inventism.
These are all imaginary isms.

The last six, have become used very frequently without reasonable purpose!
These words will come up in the following pages to show how absurd they are.

All **isms** lead to suppression, poverty, and control over people. They block human individuality and inventivity.
Take for instance: fantasism. Al Gore's theory of Carbon Footprint brought him great wealth but it accomplished nothing, it was a combination of both Inventism and fantasism.

An Anecdote

There was a man who studied at several universities, but never reached a degree.

One day he was reading about the ice cap at the North pole. It was very hot in his office.

His AC was not working very well, so he took a bowl and placed a bag of ice in it without dumping the ice out of the bag, it cooled him somewhat. When he went to bed, he forgot about the bag of ice.

The next morning when he came to his office, he found the bag of ice had melted and as it had filled the bowl the excess of water had covered his desk.

"Aha," the man exclaimed, "that is what will happen when the North pole glaciers will melt."

Simple mis-calculations are used as facts.

Zombie ice from Greenland will raise Sea levels by 3.3 %.

The Greenland ice sheet is 660.000 square miles and it averages 7500 ft in thickness.

The surface of U.S. is 3,700.000 square miles. The math tells us that is .033 %

not what the scientist told us, that it would be 3.3%.

Just a simple fact that supposedly makes rising the Oceans by 3.3 Feet.

That is just wrong mathematics!

The CO2 Scare

CO2 is rising to unacceptable levels because of the use of fossil fuels. **(Non-sense)**

CO2 has **not** caused temperatures or sea levels to rise beyond historical rates!

Severe storms have **not** increased in frequency or intensity since 1970 neither have heat waves nor droughts!

Global warming, if any at all, is **not** warming coral reefs!

Dr. Fred S. Singer, Astrophysicist, explains in his Book: "Hot Talk Cold Science":

His masterful analysis decisively shows that the pessimistic, and often alarming, global warming scenarios, depicted in the media, have no scientific basis.

The Nonsense of Global Warming and Climate Change

Arctic Oceans, Sea Ice, and Coasts

Home › Regions › Alaska and the Arctic › Arctic Oceans, Sea Ice, and Coasts ›

The Arctic Ocean is blanketed by seasonal sea ice that expands during the frigid Arctic winter, reaching a maximum average extent each March. Sea ice retreats during the Northern Hemisphere's summer, reaching its minimum extent for the year every September. Arctic ice cover plays an important role in maintaining Earth's temperature—the shiny white ice reflects light and the net heat that the ocean would otherwise absorb, keeping the Northern Hemisphere cool.

Arctic sea ice is declining at an increasing rate in all months of the year, with a stronger decline in summer months.

Arctic sea ice extent in September 2012 was the lowest in the satellite record (since 1979). The magenta line indicates the September average ice extent from 1981 to 2010.

A former Vice President who is a Fantascientist, received the Nobel Peace Prize for his efforts to obtain and disseminate information about the climate challenge.

Having no science degrees on environmental issues, he took any report on climate change for granted and created a report on CO2 and how to lower it, by selling "climate offset" certificates.

Al Gore made a fortune selling these: "Climate offsets."

The bottom line is, that after decades of "Al Gore science" **and Climate Alarmism,** we soon may no longer be dreaming about a white Christmas, instead we would be dreaming of a green Memorial Day as CO2 is no longer an issue because, as Dr. Patrick Moore wrote in 2005 that climate alarmism is preposterous and CO2 is actually plant food. In fact: Dr. Singer found that many aspects of increased CO2 levels as well as any modest warming, such as a longer growing season for food and a need to uses fossil fuels for heating, would have a highly **positive** impact on the human race.

Dr. Singer also notes how many proposed "solutions" to a Global Warming Crisis, like **"carbon" taxes** would have severe consequences for economically disadvantaged groups and nations. As alarmists clamor to impose draconian government restrictions on entire populations in order to combat "climate Change," Dr. Singer's book reveals some startling, stubbornly contradictory facts.

Higher CO2 levels greatly increase plant and crop yields.

During the Dinosaurs' age, supposedly, CO2 levels could have been 5 to 10 times higher. (in provable levels).

That is why the Mesozoic Era was characterized by lush foliage which ultimately led to our present time fossil fuels.

The bottom line is, that after decades of Al Gore **"Climate Alarmism,"** we soon may no longer have to worry about a White Christmas.

Climates are Cyclical.

An Astrobiologist, Jack O' Malley-James, stated in 2013 that the existence of the Earth would ultimately end, because of a **Shortage of CO2** which would take 1.000.000.000 years. (One Billion)!

A Danish Statistician, Bjorn Lomborg calculated that reducing the global temperature by only one third of a decree, meaning: postponing global warming in less than 4 years, would cost 100 trillion dollars.

Again, the earth is self-regulating and **man cannot change anything** of the earth.

"Global Warming," by Alex Bruesewitz.
The left has been spreading **climate misinformation** for 50 years.

Here are some dates that have long gone by, without any even the smallest changes.
1976: Global Cooling.
1989: Rising Sea Levels Will Obliterate Nations if nothing is done by 2000.
2000: Children will not know what snow is.
2002: Peak oil in 2020.
2008: Al Gore predicts Ice-free Arctic by 2013.
2023: Davos. Al Gore rants about Boiling Oceans.

Utter NONSENSE
In the Column of "Cold Science" 2021 Mike Hofman said: "Those in line to make Millions by buying "Carbon Offsets," are the most ignorant people on the earth," it made Al Gore very rich.

Intelligent and informed people do not deny that there have been thousands of climate changes over billions of years.

However, there is **no scientific evidence that man causes global warming.**

It is all computer generated, using false data. Man has never caused climate change.

Mike Hofman has been researching for 38 years, he is an engineer who studied Physics, Chemistry, Thermodynamics, heat transfer and black body radiation and absorption etc.

The Nonsense of Global Warming and Climate Change

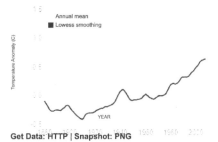

In this page of "**Nasa's Goddard Institute for Space Studies,**" it showed that from 1884 to 2021 the average global temperature rose (as an anomaly by just C 1 **degree.** That was a period of 137 years.

We can hardly call that a Hot or Heat increase!

Yet, at the 2023 annual meeting of the World Economic Forum in Davos, Al Gore ranted, in a heated speech, (no pun intended) about "**boiling Oceans**" and millions of global deaths due to global warming, calling them "**Climate Refugees",** the Forum accepted his unscientific rants as facts, and gave him a standing ovation.

AL Gore has never been right.

Real Nonsense!

Global warming does have a **temporary** source which is caused by The Earth itself:

by Volcanic Eruptions, 79 eruptions in 2021, 48 eruptions in 2022, and in 2023, 51 eruptions so far. Notice those numbers! Hardly any significant changes!

The heat from these eruptions rises upwards, but as it goes up, it meets colder atmospheres, by the time it passes the Troposphere up to 10 miles it is completely cooled down.

What "scientist" say about **Global Warming**: it is a gradual, long-term, very low increase in the average temperature of the earth's atmosphere due to the greenhouse effect, where gasses from various human activities, including the burning of fossil fuels, trap heat from solar radiation.

Again, this is **"Fantascience"** because globally the earth is cyclical and there have been warming and cooling events for as long as we have been able to record it, and many centuries before that.

Non-sensical "evidence," has been a topic of debate for centuries.

According to the American Institute of Physics, as early as 1200 BC ancient Greeks argued whether draining swamps and clearing forests might impact more rainfall.

However, many scientists worldwide, agree that Global Warming is real and that it could result in devastating effects for humanity. They all seem to ignore that we live on **a cyclical, self-correcting Earth.**

Now, today, more scientists are objecting to the "facts."

Greenhouse gasses. Believers of Global Warming have been typically attributing over-accumulation of gasses such as carbon dioxide (CO_2), nitrous oxide (N_2O), and methane (CH_4) in the Earth's atmosphere.

Sources of these gasses.

Burning Fossil Fuels: Machinery, Cars, Trucks and Trains that use coal, natural gas, or oil, release CO_2, (Carbon Dioxide) which is called Greenhouse gas, into the Atmosphere. Not to be confused with the Stratosphere.

First, CO_2 is heavier than Oxygen and CO_2 is subject to the Earth's gravity, **but "IF" it rises** upwards, it would travel through cold air which makes it even heavier! So, it could never reach the Stratosphere where the Ozone supposedly heats up these gasses which can never arrive there!

Secondly, CO_2 is absorbed by all the trees and plants on the Earth, as well as by the Oceans.

In return, all the growing plants and trees emit Oxygen. Without CO_2 we would run short of oxygen.

Remember: Our Earth is self-supporting and it is supplying all we need.

The net effect of burning fossil fuels is warming, according to some scientists, because the cooling is small compared to the heating, caused by the greenhouse effect. However, the greenhouse gases, which cause the supposed warming, remain in the atmosphere for many decades and even up to hundreds of years.

They mix with each other and spread throughout the stratosphere all around the globe and lose their strength and become low in temperature, these greenhouse gasses spread into the huge cubic miles of the universe, Like: (not a drop in a bucket), but a drop in the Ocean.

But since greenhouse gases remain in the atmosphere much longer, they will encounter mayor cooling seasons, winter, snow, rain, and hail would cause those greenhouse gases to dissolve or crystalize and gradually return onto the surface of the earth.

Many scientists scoff at the idea of global warming and believe **if** it is even occurring!

Here is a real Fact:

NOAA (National Oceanic Atmosphere Administration) reported that the average surface temperature on Earth has risen 0.74% Celsius or 1.3% Fahrenheit since the 1970th's **That is during a time span of 50 years**! But in reality, it has risen only C.013% or F0.23% per decade during the past 50 years.

Hardly noticeable raises!

Another global warming myth.

De-forestation, farming and livestock are supposed to cause global warming.

Since the 1700dreds people began to think about what would happen if The Earth would get too hot, they did not have the information or details, so they took any form of warming as a possible reason, for instance:

Cows' farting, which makes heat and is even flammable!
How many cows are on the Earth?
One and a half billion! But there are many other animals who also fart!
Is that a reason to eliminate all the cows?
How many people are on the Earth?
Almost 8 billion and they also fart!

Here is a little humor: The avoid Fart gasses, first kill all the cows, then leave it to Al Gore, he will let people buy Offsets for farting, pay him $ 100 and you may freely fart for a year.

Or, the Government should create **farting stations** and thus create a gas on which cars can drive.

More Nonsense!

The smell in the stables was a reason, the smell of manure became a reason.

Sweaty people could be a reason and when the steam engine was invented, the whole earth was going to burn up. People proclaimed!

In the 1800dreds chemistry became a reason. When the first cars were invented, exhaust gasses from cars were examined and CO2 (Carbon Oxide) became a suspect followed by N03(Nitrates) N20 (Nitrous Oxide) and CH4 (Methane) became really bad because it smelled so bad and it made smoke.

In the 1973th, the catalytic converter was invented. It changes the harmful exhaust gases of engines into safe gases before they enter the air.

Even airplanes are now being equipped with catalytic converters.

Many similar gas converters are created for Factories, Refineries and Coal plants.

We need to eliminate all the cows in the country, Ha, Ha.

NONSENSE

Why should we get rid of the cows and what would be the result and the consequences?

According to AOC, cows are harmful to society, because they smell, and they create Methane gas.

First of all: smell is not polluting, and methane gas is readily absorbed by the Earth as the nitrates from Methane is fertilizer for the whole Green Earth.

Look at the trees, they are darker green than ever!

Besides slaughtering cows is inhumane? All that blood is horrible? We do not need to eat meat?

Do understand, that **all the above reasons would hinder our lives on the Earth.**

Because, not only millions of jobs would be lost if cows would be eliminated, hundreds of products would be disappearing, to name just a few: Leather, blood meal, bone meal, glue, meat, and many, many more products which are essential to our lives.

Chief benefits of Global Warming.

Fewer winter deaths, lower energy costs, better agricultural results, because of fertilization effects of returning gasses lake CO_2 and NO_3.

Fewer droughts, richer bio-diversity, more plant and tree growths.

Milder climates, former untappable gas and oil reserves can be made available.

History teaches us more of the Truth.

Since the 1750's the average of **warming and energy remained constant**!

If a more active sun caused the global warming, scientists would have expected warmer temperatures in all the layers of the atmosphere but, because the Greenhouse Gasses are losing their heat in the different layers of the atmosphere, they **cannot cause global warming.**

Conclusion, all reasoning on global warming is utter **NONSENSE**

Here are some **facts from Scientist about the NONSENSE of Global Warming.**

In a book by Professor Davis Dilley: "From Nice Age to Ice Age" the Professor announced:

Drastic Global Cool-down predicted. Mini Ice Age coming before 2030.

However: the earth has not warmed, nor cooled in 19 years.

He said: "this is likely the reason why leftist's intent on taxing in the name of global warming," had to rename changes in Earth's climate to" Climate Change".

Over 400 Prominent Scientists disputed: "Man-Made Global Warming claims in 2007".

Climate Science's Myth-Buster

It is time to be scientific about Global Warming, says Climatologist Judith Curry.

We've all come across the images of polar bears drifting on ice floes: emblematic victims of the supposed Global Warming which is melting the polar ice caps, symbols of the threat to the Earth posed by our ceaseless energy production.

Above all," about the carbon dioxide that factories and automobiles emit, we hear louder and louder, demands to impose limits, to change our wasteful ways, to save not only the polar bears but, also the planet and ourselves". By Guy Sorman.

A Climatologists Turn-around.
In 2010 Dr Judith Curry wrote a book: "Climate Uncertainty."

Climatologist Dr. Judith Curry: "Scientists trying to hide the decline in Global Temperature," is perhaps the most infamous example of this, it comes from the "hide the decline" email.

This email initially garnered widespread media attention, as well as significant disagreement over its implications. In our view, the email, as well as the contextual history behind it, appears to show several scientists eager to present a particular viewpoint that anthropogenic emissions are largely responsible for global warming even when the data showed something different.

She took a different look at the Time Frames and the prevailing uncertainty. In 2013 she wrote: "If Climate Scientists would acknowledge the inherent uncertainty and admit that, we do not know the future facts, they should be more sensitive to their feelings and opinions.

They violated The Norms of Science.
Calling it Climate Gate! While the Climate norms before 1985 was still Shaddy.

Dr. Patrick Moor wrote in 2005: "Climate Change is Preposterous", called out that Scientists were Hiding the Nonsensical Changes."

Dr. Judith Curry now states: "There is no emergency."

Jorden Peterson wrote: "Non Alarmists Approach."

The official U.S. "Annual Climate Normal," show: from 1901 to 2020 there was a temperature rise of **Fahrenheit 1 single degree! A world-wide average".**

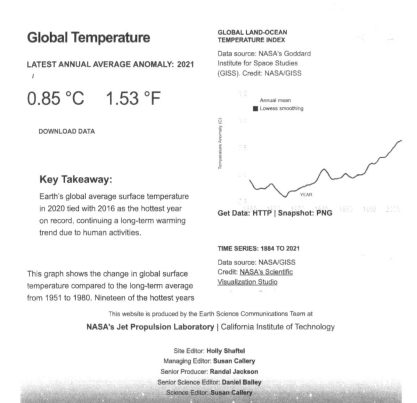

Dr. Patrick Moore wrote in 2015 **that climate alarmism is preposterous** and that CO2 is not acting the way the scientist treat it and it is essential to plant and tree growth.

In the dinosaur's age he said the CO2 levels must have been 5 to 10 times higher.

Even though, there were no cars or factories in existence!

That is why the Mesozoic Era was characterized by lush foliage which ultimately led to our abundant fossil fuels.

See, that the Earth even then, took care of itself, but also of our energy future.

Politicalization Nonsense.

How much has been spent on Global Warming in the USA?
The budget for 2022 was $ 3.6 billion.
For 2021 it was $ 1.5 trillion
For 2020 it was $40 billion
The total amount which the rest of the world has spent is astronomical!
What has been accomplished with all that money?
This is Climate Zealotry in full display.
Government, Corporations, Universities, Multimedia, Powerful Millionaires, nobody has any power or influence on this Earth. They all influence The World, but they have no power over Planet Earth. They all serve The World and specifically their own world but not The Earth. Money is Power, but it cannot change The Earth.

The Paris Conference to Combat Climate Change.

In Paris, on November 30 to December 11 in 2015, The Paris Agreement was adopted.
Below the first page of a 32-page Agreement.
At the conference 175 countries had been represented.

The NONSENSE is propagated worldwide!

False Alarm

In a "Hoover Virtual Policy Briefing," By Bjorn Lomborg, who has been studying and writing about Climate Change for twenty years. Lomborg writes in an article titled: "False Alarm": How **Climate Change panic** costs us trillions but fails to fix the planet. Lomborg is one of thousands of Scientists who disagree with the climate panic theory! They all believe it is **Utter Nonsense**.

Wasted Billions.

Imagine a meeting in Paris for ten days and 175 delegations present. Many of these little countries could hardly afford the expenses of travel to Paris. Those countries could have used the hundred-thousands of dollars for better purposes in their own country, but they had to be in Paris to listen to the **Nonsense of Global Warming!**

WHAT HAS BEEN ACCOMPLISHED SINCE 2015? Nothing!

In the 32-page report of the Paris agreement, are several (solutions) but what has really been accomplished?

Carbon Footprints

Just a little explanation **on Carbon Footprint:** It is the amount of **Carbon Dioxide (CO2)** emitted due to the consumption (use) of fossil fuels by a particular person.

However, the wind blows that percentage to vacant lands and forests and thins that percentage to barely zero. The sheer size, or cubic volume, of the Atmospheric layers shows how thin the layers are.

Besides it is being used by the vegetation on the way.

Volumes of Earth's Atmospheric Layers

Given			Dia of a Sphere =	(4/3)(Pi)(radius cubed)
Diameter of Earth			(4/3)(Pi) =	4.1888
Equator		12756	KM	
Poles		12713.6	KM	
Average Diameter of Earth		12734.8	KM	
Volume of Earth		1.08E+12	Cubic KM	

Height of Atmospheric Layers Above Earth	KM	Volume of Atmospheric Layers (Cubic KM)	Percent
Troposphere	12	6.13E+09	15%
Stratosphere	50	1.95E+10	47%
Mesosphere	80	1.56E+10	38%
Total		4.13E+10	100%

Unprofessional mis-judgements!

The negative information **on misjudged CO2 effects** on the entire Earth come from the politicians, who declare that too much CO2 is bad for the planet.

In "Science Briefs," Erwin van den Burg declares that: "In the past, this cycle of CO2 production remained balanced, with Carbon outputs and Carbon absorption running relatively even. But, no one has proven that the CO2 balance has changed world-wide!

All the sheep who believe Climate Change is caused by CO2 are fools said **Martin Lewis.** a Physicist who received the 1995 Nobel Prize on an article of, "Cold Science."

New study undercuts favorite climate myth, more CO2 is good for plants.

Dana Nuccitelli: "A 16-year study found that we are at a point where more CO2 will not keep increasing plant production, but higher temperatures will decrease it. I have been on the look-out for the best arguments why CO2 is **not causing Climate Change**".

Here is the best one I have found to date along with some counter-arguments and evidence. It looks at whether **CO2 is saturated in the atmosphere** and more CO2 will now make minimal additional impact on warming. (An article in "Science-Briefs").

They all seem to forget that CO2 is subject to gravity and thus, does not rise into the atmosphere.

Paul Bedard, ("Washington Whispers") is focusing his efforts on Congress, convincing them that **CO2 emissions are great for our planet**, and **do not cause climate change.**

Plenty of NONSENSE

Distinguished Astrophysicist Dr. S. Fred Singer wrote in his book:" Hot Talk, Cold science," a masterful analysis that shows that the pessimistic, and often alarming, Global Warming scenarios depicted in the media have **no scientific basis**. He also notes, how many proposed "solutions" to the Global Warming crisis, (like **"carbon taxes"**) **would have severe consequences** for economically disadvantaged groups and nations.

As alarmists clamor to impose draconian government restrictions on entire countries, in order to **combat Climate Change,** Singer reveals some startling contradictory facts.

Such as: **CO2 has not caused temperatures or sea levels to rise** beyond any historical degree.

Severe storms have **no**t increased in frequency or intensity since 1970, neither: have heat waves **nor** droughts.

Global Change has **not** harmed any coral reefs.

Any increases in CO2 concentrations across huge time spans have **not** preceded rising global temperatures; those concentrations were followed them by some eight hundred years. Which is just the opposite of the alarmist's claims.

Conclusion: despite all the hot talk, (pun intended) and outright duplicity, there is **no Climate Crisis resulting from human activities** and **no** such threat is on the horizon.

Dr. Judith Curry's message to Capitol Hill.

While humans may be contributing to Climate Change, we simply don't know how the Climate will behave (The Earth), in the coming decades. So, there may be no point in trying to reduce emissions.

The Earth will deal with it and corrects anything in time!

Republican member Dana Rohrabacher **sees Climate Change as a Liberal Plot**!

Dr Judith Curry continues:

We have gone through warming and cooling trends, but how much if anything has to be contributed to human activity? She asked rhetorically. Concerns about Climate Change gives The Government control over the people, meaning over our lives and our freedom.

She called it the rough-and-tumble climate debate. Her philosophy then and now, is that if Climate Scientists would more readily acknowledge the **uncertainties** inherent to the issue, skeptics would more likely accept the well-established central tenants of Global Warming. This could change the ideas of the IPCC.

Dr. Curry wrote an essay on: "The credibility of Climate Science."

She wrote: There is no evidence that Climate Change is accelerating the rise of the sea level. From 1901 till 2020 the sea-level rose only one sixteenth of an inch, that was in a period of 119 years!

Once more, it proves that man cannot correct the nature of the Earth.

And here is an **inconvenient truth!**

CO2 in the Earth's atmosphere is now about 410 parts per million and it was about270 parts per million 300 years ago! If CO2 levels would get below 150 parts per million, photosynthesis would stop and every green thing on this Earth would turn brown and die and we, the people of THE WORLD would die too, from a shortage of OXIGEN.

Climate Change

Climate Change is a significant variation of weather patterns over long period, however:

David Dilley, CEO of Global Weather Oscillations Inc. gave us an inside into the Climate Pulse Cycles:

There are cycles occurring every 6 months. Every 4 years. Every 9 years. Every 18 years. Every 72 years. Every 230 years and Every 1200 years.

Buck Meadows wrote: "**Climate Change is used for A New World Order.**"

Yes, it is!

But it happened for as long as we have recorded history and many hundreds of centuries before that!

So, why then, is Climate Change more important, at this time and age?

Here are some key **Climate Change Statistic gathered by NASA:**

Over the past **two** centuries, the average global temperatures of the Earth surface have risen by 2.12 degrees Fahrenheit (1,18 degrees Celsius). Which is 1.6 degrees Fahrenheit and 0.9 degrees in Celsius in a hundred years. Break that down per year, it rose by F. 0.016 and C. 0.009. that is practically nothing!

Not even 1 thousands of a degree Celsius, and 0.16-tenth of a thousand-degree Fahrenheit!

Is that something of a reason to sell your winter coat?

In fact: The ice in the North- and South poles was breaking because it was getting so old, that it started crumbling. Resulting from wind and old age!

No mention what happened with the many billions of Tons of ice under the sea level.

Nothing!

More NONSENSE

Seventy percent of the Earth is water and of that 96.5 Percent is in the oceans.

If all the ice would melt, (which will never happen in a million years), it would possibly raise the sea levels by half an inch, because 95 Percent of that ice is already under the sea level. Thus, it cannot change the sea level.

In the past 50 years there have been several cooling periods on the globe!

That is the proof that the Earth is self-correcting and Cyclical.
Meteorologist, Author, Joe Bastardi, emphatically denies that global warming is human induced, even though Joe Biden calls it a Climate Emergency.

"The weaponization of weather," Bastardi quotes, "is a phony Climate War." (Sept.8,2020)

"It is politically incorrect" said, Marc Morano.

Climate Change Causes:

Natural causes:
1. Changes in solar radiation, Sunshine. The sun's rays heat The Earth regardless of changes in weather patterns that occur on Earth. As Such, any change in the sun's radiation, either an increase or a decrease, will influence our surface temperatures. Man **cannot** change that ever!

2. Greenhouse gasses.
The untrue Myth's:
The various gasses, created by man, supposedly rise in the air and trap heat in the Earth atmosphere and are then thinning the Ozone

layer. However, these gasses cool down as they rise higher, and thin in the ever-widening atmosphere.

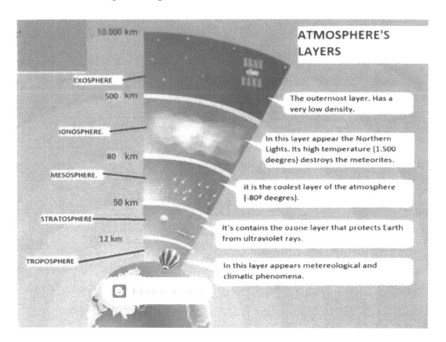

The Ozone layer protects the Earth from ultraviolet rays. **If** there is any depletion of Ozone, it has never been proven, there might be a slight variation in Ozone percentage but it is so small that it is in-measurable.

3. Drastic weather Changes.

Disasters like hurricanes, floods, tornados are natural causes which temporarily changes the climate, (for a limited short while) but these climate changes are localized and therefore have no global influence.

Man-blamed causes of Climate Change.

Industrialization

Factories which are releasing greater amounts of greenhouse gasses are blamed for Climate Change but these gasses are seldom rising into the higher atmosphere. The smoke from these factories contains mostly steam which crystalizes in to clouds and comes back to earth in the form of rain, snow, or hail. In the past twenty years, these factories have cleaned up their smoke with all kind of filters.

When those gasses go up-wards, they quickly get cold in the higher altitude.

There is no reason this could cause Climate Change.

Inconsistent Emission Controls

The Catalytic converter has mostly taken care of the exhaust gasses. This means that fossil fuels are good to use and **cannot cause Climate Change**

Deforestation myth.

This is a myth which has been used as far back as the Roman Era. In the Middle Ages, chopping a tree was a crime and even now, some cities demand people to get a permit to cut a tree.

Because trees breathe-in the CO_2, deforestation would lessen the use of CO_2, but in a few months, plants start to grow again, even after a forest fire! However, the earth makes the deforested area quickly green again, and man uses the deforested area for agriculture, any crop grown on that land uses CO_2 again.

See the cover of this book! Two years ago, these mountains burned totally, but within two years they are blooming again and growing trees quickly as well.

Here too, there is no reason for climate change!

More non-sense.

Agribusiness

All farms, all over the world create CO_2 (Carbon Di Oxide) and CH_4 (methane). But so do all animals, in the wild, as well as Humans all over the world. These gasses are released in the air and people have **blamed** that to global warming. The increasing temperatures as these gasses are supposed to be trapping more heat in the atmosphere and thinning the Ozone layer. More on Ozone below.

This theory is untruthful and foolish. The facts are; that the higher the gasses rise (if at all) the colder and thinner the layers of the atmosphere and stratosphere are and it makes all gasses cooler. There is no trapping of heat.

Moreover, CO2 is heavier than Oxygen and CO2 returns to the earth to feed all the plants and trees in this **green earth**. In turn for the Oxygen, we all breathe.

(picture of the green earth)# Foto to make

Everything the world, that is The People, eat, produce, build, manufacture, create, consume and use comes from **THE EARTH** and!!! returns to **The Earth**.

Modernization

The concrete used to build roads, and the vehicles that travel upon them used to create higher levels of CO2. Thanks to the invention of the catalytic converter those numbers of CO2 have been drastically decreased, even planes are using some form of catalytic converters.

Yet the politicians use the exhaust fumes as an excuse for Global Warming and that causes them to **blame fossil fuels.**

Global Warming and Climate Change

The two most popular terms to describe the (supposedly) earth's increasing temperatures are called "Global Warming and Climate Change".

But remember the facts about both terms, the increases are so small over the past hundred years that it would take thousands of years to become harmful to the world's inhabitants. Yet with **fear tactics**, mainly coming **from unscientific politicians,** Global Warming theories remain a threat to the people from the Earth.

Great NONSENSE

Climate Change Causes

A really small number of climate scientist agree that Climate Change is actively happening. And, when it comes to pinpointing the specific causes for those changes, there is hardly any consensus.

Some argue that natural variations in atmospheric conditions are heating the planet while others claim that humans are responsible for this increase, which is, in reality, so small that it would take thousands of years to become a threat to the earth.

Remember, the Earth is Self-Regulating and no human being can change the Earth, again and again the Earth proves it to us.

We humans can never change earth-quakes, tsunamis, volcanic eruptions hurricanes or tornadoes. They have always occurred and will always happen again.

Some **Climate Change Statistics** gathered by **NASA**.

Over the **past two hundred years** the global temperatures of the Earth's surface have risen by 2.12 degrees Fahrenheit (1.18 degrees Celsius).

Between 1993 and 2019 there was an annual loss of 279 billion tons of ice in Greenland and 148 billion tons of ice in Antarctica. Who has weighed that? How much of that ice has just sunk into the deep? How much has that risen the oceans? Not even one third of one inch!

The point is, that 95 % of frozen waters are under the sea level. Only Ice that rests on the bottom of the waters can rise above the water level, thus creating glaciers.

Remember the Titanic? It ran into an underwater iceberg. They could not see the iceberg until the ship hit it.

So, if 279 billion tons of ice in Greenland was lost, only 5% of that could have been above the water level, which also means that approximately, only 14 billion tons could raise the oceans by a fraction of an inch, based on the massive 361 million square kilometers of oceans.

It seems that with all the extensive studies by hundreds of scientists, they have forgotten that so much ice is already part of the water levels because it is already in the water and below the sea level.

The Nonsense continues!

The cost of "Going Green", AOC's hobby horse!
Xenophobia is a combination of two Greek words: which means "stranger or guest" and the word has been accepted and used as "Fear or Panic."

Politicians use Xenophobia to create fear in people on Climate Change, Global Warming, drink water scarcity, calling and blaming anything a Climate Crisis. It is also used to split the public opinion in two agendas about the world: Woke-ism and extremist on World policies, this will be explained later.

No Green Deal is harmless.
Reducing CO_2 is un-professional thinking, CO_2 is needed for our existence. Many studies have proven that! The money spent on studying what has been common knowledge for hundreds of years is a waste of time and money.

NOTE! For a New Green World **Hint.**
Paint all the buildings and roads green, because man cannot improve a newer green Earth, it is already green. Wherever there is soil, the Earth will make it green!

Red China's Lithium Time Bomb
Ignites a Lithium Stock Megatrend
Despite the bear market, inflation and recession, select lithium stocks are soaring

The U.S. is dangerously dependent on the Chinese for the lithium needed to power electric vehicles and to store the energy generated by solar power.

China produces 51% of the world's lithium and processes 72% of it.

And they could cut us off at any time.

Electricity from batteries

used for thousands of consumer products including cars, have serious consequences. Lithium and Cobalt are the ubiquitous power supplier for batteries. Cobalt ions are needed and it is considered, the highest material supply chain risk, for electric cars.

Graphite is another unreplaceable element used in Batteries.

The need can become too big, so that the Earth may not be able to supply the growing demand. Graphite is also a fossil fuel and mining are done is some twenty countries. (See map). Some companies warn that we might run out of supplies by 2035. The mining of Graphite is similar to coal mining and that has been shut down gradually in The United States.

People put their hope in Electrification and sacrificing traditional sources of energy.

That is another Nonsense if we continue the replacement theory!

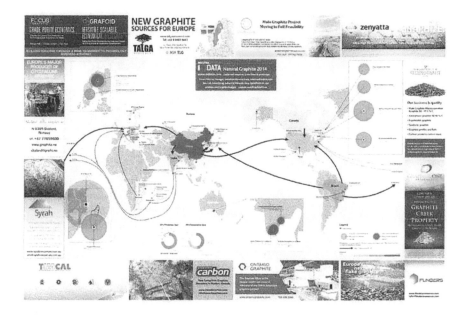

A Carbon Footprint of Lithium,

Since Lithium is used in almost all batteries, is just as serious as the use of fossil fuels. The environmental impact begins at the mining and continues well into the processing of batteries, but the disposing of the used batteries poses an even larger problem.

Because of how valuable it has become and how terrible the working conditions often are for those mining it, cobalt is known as the "blood diamond of batteries."

Check this out!

Slave labor and child abuse on a mass scale in order to drive electric cars.

Cobalt is mined in the Congo, 70 % of the World's Cobalt is mined through the exploitation of children of 6 to 12 years of age.

40.000 children work with their bare hands to separate Cobalt from broken rocks and soil at a price of 10 cents per hour, in order to enable millions of our children to play with their I-pads!

If this slavery will be stopped and the mining has to be done by adults at a minimal wage, Cobalt will become too expensive to use. That is one of the costs of a **New Green Deal.**

There are many writings, books, and Scientific reports on the New Green Deal.

Here are some outstanding books on the phony Climate War.

The weaponization of Weather in the phony Climate War," by Joe Bastardi.

The Politically Incorrect Guides," by Marc Morano.

Climate Crisis and The Global Green New Deal," by Noam Chomsky and Robert Pollin.

The origin and details of The Green New Deal!

The term "Green New Deal" was first used by Pulitzer Prize-winner Thomas Friedman, in January 2007. At that point America experienced its hottest year on record (even though) there were five hotter periods before that.

Friedman recognized the easy solution to climate change the politicians hoped for was not really possible.

The January 2007 announcement was made popular by a proposal of Rep. Alexandria Ocasio-Cortez in Congress in 2019.

The deal, which failed to pass in the Senate, emphasizes environmental and social justice while creating new jobs.

Supporters said: everyone is responsible to pay their fair share, which would result in tremendous cost savings.

However, implementing the deal would cost at least as much as $ 93 trillion. One of those "solutions" would be to transition away from fossil fuels. The Government would require to raise prices on them, introduce higher energy standards and undertake a massive industrial project to scale up green technology.

The plan was introduced by un-knowledgeable politicians, who had no idea of the tremendously utilization of fossil fuels.

The Green New Deal has become a swamp of rules, regulations, un-attainable ideas, resulting in an unbelievable green alphabet as follows:

Agro-ecology, Carbon fee and dividend, Carbon Finance, Carbon negative fuel, Circular economy, Earth economy, Eco-capitalism, Eco commerce, Economics of global warming, Eco-metrics, Eco-socialism, Eco Money, Ecosystems, Ecotax, Energy balance, Environmental accounting, Environmental economists, Environmental credit crunch, Environmental enterprise, Environmental investment, Environmental pricing reform, Environmental tariff, Fair Trade, Fiscal environmentalism, free-market environmentalism, Green banking, Green libertarianism, Green syndicalism, Green trading, Gross domestic product, Natural capital, Natural resource, Natural resource economics, Principles of eco-preneurship, Property eco-rights, Renewable resources, Eco-risk assessment, Strategic Sustainable Investing, Systems ecology, World Ecological Forum.

They are all Hypotheses and Theoretisms, are leading to hundreds of Committees, meetings and reports, which all have hardly any sense nor value. In other words: **Hypothetical Nonsense**. Imagine the expenses resulting from it!

How would the Green New Deal create jobs?

The Green New Deal promises to create millions of jobs, largely in the Governments by tackling economic inequality. We would be "guaranteed" high quality jobs backed by Unions! (?) by shifting money from the fossil fuel industry to green technology. The deal supports the inclusion of traditionally marginalized individuals such as (illegal immigrants) indigenous and racial diverse communities.

What a nonsense!

What is in the Green New Deal?

It is all about changing **THE Earth** by putting **The World** in charge of the environment, by changing the weather, (which no one ever can) and in changing the way we all live!

The Green New Deal is supposed the create millions of new jobs, for access to nature, clean air and fresh water, healthy food, a sustainable environment, and community resiliency. These goals are supposed to be accomplished through the following actions by the Federal Government:

Providing **investments** and leveraging funding to help communities affected by climate change. **(Impossible)**

Repairing and upgrading existing **infrastructure** to withstand extreme weather and ensuring all bills related to infrastructure in Congress address climate change. **(Impossible)**

Investing in Manufacturing and industry to spur growth in the use of clean energy. (**Just clean up our existing energy**)

Identifying unknown sources of **pollution and emissions. (Has already taken place)** needed to turn the Green New Deal into a reality. (More power to the Government that is what they are after!)

Eliminating Fossil Fuels and the consequences!

Please understand that fossil fuels run the world! Life as we live it would come to a standstill if we ran out of these important compounds.

Fossil Fuels supplement the production of hundreds, if not thousands, daily used products such as household items, clothes, cars, airplanes, boats, pesticides, wrapping materials, bags, purses, and a lot of food is preserved with a little help from fossil fuels.

Yes, you read that right. The food we eat is preserved using fossil fuels!

Natural gas, coal and oil are examples of fossil fuels. The name "fossil" denotes a substance produced from dead decaying matter accumulated over millions of years.

Fossil fuels were formed more than 650 million years ago. The energy was released from carbon stored in dead matter. The energy from the decaying process was converting into natural gas, coal, and oil. Fossil fuels are a non-renewable source of energy.

Petroleum, Kerosene, Methane, Propane, Butane, Coal, and many more are examples of the fossil fuel products. They are all a source of many different raw materials used directly and as raw-material for the manufacturing of countless products for our daily life.

Following are 75 common uses of fossil fuels.

1. Fuel

The most common use of fossil fuel is gasoline and diesel. But that is not the only use. Out of 43 barrels of oil, only 19.5 gallons are used to produce gasoline.

Almost 60% of crude oil goes into manufacturing different products, the bulk of which are plastics.

Gasoline, Jet Fuel, Diesel, Heating oil, Kerosine are processed from crude oil. Both Petroleum and Natural gas are then used to produce electricity which powers industries and lights homes. (In total hundreds of products.)

2. In cars,

Cars are made of about 50% plastics. Almost all plastic components in cars are made from petroleum by-products. Plastics are versatile, durable, lightweight, and easy to replace. This makes them ideal raw materials for, car handles, air vents, dashboards, airbags, doors, and fenders. (**Totaling into hundreds of items.**)

3. Petroleum Jelly.

We all know Vaseline, a worldwide brand of petroleum jelly. Used as a skin toning application. Petroleum jelly is made from crude oil and used in hundreds of creams and ointments. The crude material is vacuum distilled, filtered through bone char to produce the jelly. It is

an odorless semi-solid mixture of hydrocarbons. **That is CO2! (More hundreds of products!)**
Who would want to live without all those creams and ointments?

4. Toys

Toys are made from blow molding plastics. Some tots such as head parts of girl's dolls and kick balls are made from rotational molding, which creates a seamless finish. Thousands of toys are made from plastics, would you tell your children they cannot have toys anymore? **(Thousands of products lost?)**

5. Computers

All plastic parts in a computer are made from fossil fuels. The computer's insulation and parts that protect a computer from overheating, including capacitors, and electrical components which include polymer capacitors are made from tetracyanoquinodimethane (C12H4N4) which is a fossil fuel product. Notice the methane part of that long word! **(Who would want to live without a computer).**

6. Asphalt and Bitumen

Asphalt or bitumen as it is commonly referred to, is a heavy black substance, a byproduct of petroleum. It is a strong, versatile, weather-resistant, binding material used in paving roads, but also for roofing houses. It binds itself well with gravel, sand, and cement to form a tough road surface.
(Would anyone want to go back to dirt roads?)

7. Synthetic rubber

Synthetic rubber is an artificial elastomer made from synthesized polymers of petroleum by-products. It is used in making doors, window profiles, water hoses, belts, matting, flooring and many other products, two-thirds of material used in making car- tires are synthetic rubber.
(We could not live without the hundreds of these products.)

8. Paraffin Wax

Paraffin wax is obtained from dewaxing petroleum. It is used in the manufacturing of candles, wax paper, shoe polishes, cosmetics, electrical insulators, and many other products. (Stop shining your car!)

(Another hundreds of products we could not live without.)

9. Fertilizers

Fertilizers used in improving soil fertility are produced synthetically. They include natural gas, a fossil fuel! Ammonia, Nitrogen, and Potassium are synthesized from fossil fuels. Phosphorus is made using Sulphur and Phosphate rock. (All farmers use these fertilizers.)

Food would become very expensive without these natural fertilizers, if there were no fossil fuels.

10. Pesticides

Pesticides help keep unwanted pests under control. They are hydrocarbons synthesized in a laboratory. Chlorine, Oxygen, Sulfur, Phosphorus, Nitrogen, and Bromine are commonly used components. Inert ingredients are dependent on pesticide type, liquid pesticides use kerosene while others use petroleum distillate as a carrier. People generally regard pesticides as poisons, because they are used to kill insects. They are mostly not poisonous to people.

Would you rather eat the insects?

11. Detergents

Detergents used in daily life, households and commercial cleaning chores are made from saponification. That is a process which involves heating fats and oils, which then react with alkali and glycerin. Hydrocarbons found in petroleum and oils are an important component in making soap, which then repels water while attracting oil and grease. Who would not want to live without soaps.

(Another hundreds of fossil fuels use we cannot live without.)

12. Furniture

Traditional furniture was made of wood. Modern furniture is made from a variety of material including metal, and plastics. Consequently, plastics are made from hydrocarbons monomers, for example, styrene vinyl chloride, acrylonitrile are used in the fabrication of high-end furniture. All derived from fossil fuels.

Who would like to go back to wood or steel furniture?

13. Packaging materials

Plastic packaging materials keep products especially foodstuff fresh and well protected. They are made from hydrocarbons, cellulose, coal, and natural gas. There is no product that can replace these products.

Unless back to using used newspapers again. we want to go

14. Surfboards and many sporting equipment's

Surfboards are made from foam they are durable and versatile. Foam is processed from polyurethane and encased within a polyester resin. More modern novice surfboards now use an epoxy resin and prolapse polystyrene (PPS) foam, rather than the polyurethane and polyester. All surfboards use fiberglass. Many sporting items, such as skate boards and roller blades are made from the same materials. All originating from fossil fuels.

Who wants to keep his kids from sporting?

15. Paints

Pains are made from both organic and inorganic pigments. Most pigments used today are inorganic. Synthetic organic pigments are derived from coal tar and other petrochemical products. They are shinier, have a better gloss and produce a beautiful finish.

Here is an example that coal and coal products are essential to our daily life.

16. Artificial fibers

Artificial fibers are made from petrochemicals. Two compounds derived from petroleum are polymerized to form a chemical bond that produces adjacent carbon atoms. Different chemical compounds are used to produce different types of synthetic fibers.

Here again, our lives depend on carbons, but the governments want to eliminate all carbons.

17. Upholstery

Car upholstery consists of fabric, padding, webbing, and springs. Foam and other hydrocarbon products are preferred material because of their versatility and durability. The foam makes comfortable padding and does not wear and tear easily.

Who would rather exchange the interior of their car, made of wood again!

18. Carpets

Carpets are made from fibers; natural fibers include wool and flax. Modern carpets are made from synthetic fiber. Synthetic fibers like nylon, polypropylene and polyester are all produced by the same chemical processes using oil and natural gas.

There is no end to the range of products originating from fossil fuels.

It is impossible to replace all the thousands of products made from oil, natural gas, and coal.

19. Solvents Diesel Motor, Oil Bearing

Solvent products made from petroleum are used to reduce friction and wear between bearing metallic surfaces. They are made by refining a solvent with a hydrogen treatment to remove non-hydrocarbons.

Petroleum helps us from having to breath in metallic dust which would come from the friction which occurs in engines of the cars we drive!

That would be real pollution.

20. Floor wax and car wax

Floor wax, used to shine wooden floors, is made from paraffin wax which is a hydrocarbon fuel which is obtained by dewaxing light lubricating oil stocks, the same goes for car wax.

We could possibly replace these waxes with cow or pig fats, **but everyone would be slipping and sliding all over the place!**

21. Ballpoint pens

Ballpoint pens are made of plastic obtained from hydrocarbons. The ink is a mixture of water-resistant synthetic pigments from fossil fuels.

Let's go back to pre-historic pencils made of wood with charred sticks. Or from graphite which was found in the 1400s. Graphite is one of three natural forms of pure carbon which had to be mined that long ago. The other pure carbons are Coal and diamonds.

Imagine missing diamonds.

22. Insecticides

Insecticides help get rid of pesky bugs. They are synthesized in the laboratory from hydrocarbons derived from petroleum. They contain Chlorine, Oxygen, Sulfur, Phosphorus, Nitrogen and Bromine. Inert ingredients depend on the type of pesticides. Liquid pesticides use kerosene or other petroleum distillates as a carrier.

Organic produce, grown in the fields are not sprayed with these insecticides to prevents us from eating bugs. However, they are still fertilized with fossil fuel-based fertilizers!

Organic food is a big lie.

23. Tires

Synthetic rubber made from polymers found in crude oil is used to make car tires. Carbon black is a fine, soft powder created when crude oil or natural gas is burned with a limited amount of oxygen, causing incomplete combustion, and creating a large amount of fine soot that is synthesized to rubber.

Sports car Bodies must be light and strong. Plastic parts made from petroleum make up almost 50% of a sports car. Seats, dashboards, bumpers, and engine components are all made of plastic or synthesized rubber.

We would have to give up car races if we did not use fossil fuels!

24. Dresses

Many dresses are made from polymer fibers. Polymer fibers are soft, warm, have a wool texture and are light in weight. They are easy to dye and hold colorfast. With all these benefits you can understand why the garment industry leans heavily on petroleum-derived synthetic fibers for colorful dresses.

Could we imagine without these fossil fuel derived products.

Would anyone want to go back to burlap clothing?

25. Perfumes, and after-shaves

Perfumes and after-shaves are made from organic plant or animal oils, but they must be dissolved in a solvent. Petroleum, ether, toluene, and benzene are used to extract fresh perfume oils from plants like jasmine, mimosa, lavender, sandalwood and many more. Once the extraction is completed, the petroleum-based solvent evaporates, bringing CO_2 in to the air we breathe.

Without the fossil fuel-based solvents our perfumes etcetera, would become too greasy!

Following are 50 more groups of products which each are a major part of our daily life. Each group is made up of hundreds of items we all use every day:

Tetra Pak, Kayaks, Toothpaste, Hair spray, Boats, Sweaters, Football Cleats, Cassettes Dishwasher soaps, Computers, Shoe polish, Nail polish, Bicycles, Motorcycle and Bicycle Helmets, Jelly transparent tape, Compacts discs, Antiseptics, Cloth lines, Curtains, Food Preservatives, Basketballs, Fishing Lures, Golf Bags, Soaps, Anesthetics, Crumb Rubber, Body Lotions, Face Creams, Toothpaste, Deodorants, Crayons, Pantyhose, Chewing Gum, Dentures, Sanitary Products, Contact Lenses, Refrigerators, Nonstick Pots, Medicines, Solar Panels, Styrofoam, Cushions, Linoleum Tiles, Toilet Seats, Rugs,

Shampoo, Tennis Rackets, Heart Valves, Surgeons Scrubs, Balloons, Hot air Balloons, Airplanes.

The list is endless and the number of products runs into the numberless thousands. Fossil fuels as you have seen, really run the world. Imagine for a moment a world without Petrochemicals, Hydro Carbons, and Natural Gas and Coal.

How would we produce the common everyday products that we are taking for granted?

Author Alex Epstein wrote in his book: "The Moral Case for Fossil Fuels"

Fossil Fuels are good for humanity, the energy available in Fossil Fuels is **indispensable** of people. For the flourishing of mankind, curtailing that energy source will bring death to billions.

Drinkwater scarcity?

As a Horticulturist, I, the writer of this book, Dr. Dirk van Leenen, having travelled the entire world, have been faced with fresh water needs for horticulture as well as for human consumption. **We will never, ever! run out of water** in the United States. Nor on the entire Earth! Remember, water is 69 % of the Earth.

We are surrounded by huge oceans, and with some effort and adequate money, we can build modern day, de-salination plants on every side of the country. We can pipe that water all over the country and never run out!

Those de-salination plants can virtually run for free once we use Solar and wind energy. Interestingly, those plants have several by-products which can be used for agriculture and industries. Even if the cost of de-salination would go up, fresh water would still be cheap enough. Our golf courses would remain green and our world would remain as-green-as-ever.

These plants have been built in several mid-east countries, including Israel. Those countries are now, and remain a "green world", there is nothing "new" about those green world countries.

But, is it true that the oceans are becoming acidic?

Many scientists have concentrated their studies on the "Acidification of the Oceans."

The sheer thought of that happening in one of the most theoretical fantasies. The reasons behind these fantasies are, that the water warms up, and before ice can form again in the fall and the winter, the ocean must release some of that heat to the atmosphere. Scientist are concerned that this increased heat transfer to the atmosphere could magnify future climatic warning trends.

This is why Al Gore was ranting about Boiling Oceans in Davos

Click this caption to open a time-lapse video of coastal change along Alaska's North Slope. The video captures action on the coast of Barter Island in Alaska over three summer months in 2014. Images show disappearance of pack ice and subsequent impact to the beach and cliffs from storms.

Ocean acidification

(This is beyond fantasy because the heat they are talking about might be as little as half a degree of Fahrenheit, and by the time is reaches the atmosphere it will have turned into zero below.)

More Nonsense

As human emissions of carbon dioxide from the burning of fossil fuels increase, the ocean absorbs more carbon dioxide and the PH of its waters decrease, making the ocean more acidic, Scientist claim.

The present Ph is 8.1, which is quite alkaline as 7 is neutral, so under seven begins to be acidic.

Remember the size of the Oceans in the Earth? **Is it more than 139 million square miles! Even a few degrees of heat could not possibly raise the oceans temperature, nor make the oceans acidic.**

The Scientists call it; a phenomenon of ocean acidification!

Research shows that ocean acidification is already affecting northern waters That very little warming of (by the way salt water) gives the water a corrosive quality that interferes with the ability of key species of shellfish to grow their shells. (Poor shellfish).

More Nonsense.

Researchers tell us that the polar ocean is particularly prone to acidification because of the low temperature and low salt content, the latter **supposedly** resulting from the large freshwater input from melting sea ice and from large rivers.

This has been happening for thousands of years! Suddenly it is a problem now?

The scientists do not even know!

The impacts on the food chain are not yet fully understood, but already some of the essential plankton and small invertebrates are showing physical changes. Ocean acidification may cause a loss of primary food sources for species that are physically, economically, and culturally essential.

They are all forgetting that the earth is Self-correcting and Self-supporting.

The following are some of the Case Studies on this subject:
Assessing the Timing of Coastal Changes in Western Alaska>
Building Resilience in the Face of Ocean Acidification>
Defending in Place: Shaktoolik's Adaptation Plan Supports Local Decision Making>
Developing Monitoring Programs for Protecting Lands in Alaska>
Looking to the future on Alaska's North Slope>

All these case studies are a waste of money, because The Earth is Self-Correcting.

Here is an example of it:

Shrubs are transforming the face of the arctic tundra. Alder, Willows, and dwarf Birch have recently moved into areas where they never used to grow. It appears that climatic warming in boreal forests has pushed many trees beyond the limits of their optimal growing condition.

Because the earth is self-adjusting!

Pollutions

Definition of pollutants

Pollution is the introduction of **harmful materials** into the environments.

They can be natural, such as volcanic ash. They can be created by human activity, such as trash or from smoke created by factories.

We can answer many questions that come from the word Trash. Every household creates trash, but is that harmful? Trash goes to a dump and is either burned or buried. In the case of burning trash, that will create smoke which could be toxic, such as Methane (CO4), but it is not toxic for long, as soon as it gets into the higher atmosphere it gets very cold and then changes into NO2 then it comes back to earth in the form as Nitrogen fertilizer. In the United Kingdom the regulations to clean up the smoke of factories has been greatly advanced by means of "Wet-Scrubbers."

Here are the different types of Pollution:
Air Pollution.
Water Pollution.
Land Pollution.

Many things that are useful to people produce Pollution.

Cars spew pollutants from their exhaust pipes, but most of that **has been resolved** with the use of catalytic converters. Burning coal to create electricity pollutes the air, but that **can be resolved** as it has been in several other countries. Industries and homes create garbage and sewage that can pollute land and water. But many Cities have sewage plants that **clean the sewage and generate gas** which is used by the cities' vehicles. The Garbage is returned to the Earth by means of **garbage dumps**.

In time, garbage dumps are a very good way to return everything "which we call waste" to the earth. When digging into a 50-year-old dump, we can find mainly dirt, even plastic and steel breaks down into powder and eventually the entire dump become useable soil. Older dumps are turned into beautiful parks.

Pesticides, chemical poisons are used to kill weeds and insects, they seep into waterways and can harm fish and wildlife.

As good stewards of the Earth we can easily **create pesticides out of natural products,** this is already happening and many produce products are treated with those natural insect deterrents, growing organic fruits and vegetables.

Pollution is a global problem, but is does not have to be a problem, it is not a matter of **Nonsense** however, since solutions are available to clean up the problems. Globally the problem could be solved if undeveloped countries, China, India, and many other countries would follow the rules made by the international conventions.

Air pollution
Peoples and Governments have been responding very well to reduce air pollution.

Chemicals called chlorofluorocarbons (CFC) have been a dangerous form of air pollution and governments have worked to reduce (CFC) in the 1980s and 1990s by changing certain refrigerants into non-producing (CFC) refrigerants and aerosols.

The **Nonsense** is in the theory of CFCs damaging the Ozone layer. That is explained in the chapter on "The Ozone layer," in which I explained that these gasses never reach that far into the Ozone layer.

All these normal living situations are blamed to cause Global Warming and that is the greatest **NONSENSE** of all. The Earth has always been adjusting to these pollutions and **Climate Change** and

Global warming, which have been happening for as long as the Earth exists have never been caused by pollution. Period!

Natural Gas is a Fossil Fuel too.

70% of cooking in homes and restaurants is done with natural gas.

The stove can produce a little CO_2 and formaldehyde which could be toxic to people and pets. For that reason, gas stoves have an Exhausts Hood above the stove, but when these do not work negative gasses can harm people's health. So, in order to solve that, we must make sure the Exhaust Hood works properly.

Is that a reason to ban all gas stoves?

It would ruin an entire industry and a loss of hundreds of thousands of Jobs and cost billions of dollars on replacing gas stoves.

What a Nonsense!

Water Pollution

There are two major types of water pollution.

One kind is visible and the other kind in invisible because it is chemical.

Visible water pollution is simply garbage, it is called **Aquatic Trash**.

Most of this trash comes from land-based activities it is human littering and it can be avoided by the people. There is also a lot of illegal dumping from marine activity and boaters as well as beach visitors. Often there is insufficient disposal equipment, but lots of people do not clean up after their stay. Even simple cigarette buds that contain plastics that will remain in the environment for many years.

According to the National Oceanic and Atmospheric Administration (NOAA), Marine debris is defined as any persistent solid material that is manufactured or processed and directly or indirectly, intentionally or unintentionally disposed of or abandoned into the rivers, lakes or oceans will become marine debris.

NOAA's Marine Debris Program is addressing that problem!

Impacts of Aquatic Trash

It affects water quality and can endanger plants and animals and it pollutes the great outdoors that we depend on for recreation. That kind of pollution is an eyesore to humanity, but for the most part it is the plastics that have a long-term negative effect. But there is a part of it that fits in the **Nonsense** category.

The main problem with plastics is that it takes many years to become micro plastics and even though it is eventually returning to dust on the Earth and in the oceans. It is **not** affecting Global Warming or Climate Change.

Here is a new invention, by Kamala Harris on 3-10-2024

Children's health is affected, she called it "Climate Mental Health."

Supposedly, the children became weary of the changing climate and are becoming confused and mentally disturbed.

How much more **nonsence** does it take to lose our mind?

One more form of Pollution and how the Earth takes care of it.

The Great Pacific Garbage Patch

It is also known as the Pacific Vortex. The size of it is 1.6 million square kilometers located in the Ocean between Hawaii and California. Twice the size of Texas.

It contains mostly plastics, which is broken down to Micro plastics. Eventually it sinks to the bottom. Plastics are made from Fossil Fuels and this is another form of "Dust to Dust" this "dust" can be harmful to marine life.

Blaming the environment.

Some Scientist try to blame its existence to global warming because the sun reflects heat on the debris which would rise in the atmosphere. Another reason for **Nonsense,** because the cold ocean water and the air above it cools that heat very quickly. This Garbage Pach is another sign of The Earth at work, gathering unsightly material into one place, far-far away from land for humans invisible.

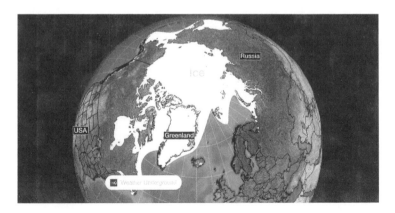

List of Sources and Quotations

Adam Curry, "No Agenda Show" The Netherlands.
Al Gore, "An Inconvenient Truth."
Alex Epstein, "The Moral Case for Fossil Fuel."
Alex Bruesewitz, Conservative Political consultant.
Arthur E Kennelly, Physicist.
Bjorn Lomborg, "The Skeptical Environmentalist" and "False Alarm."
Buck Meadows, "Climate Change is used for A New World Order."
Carl Friedrich Gauss. Speculated in 1839 on Variations in the Earth's Magnetic Field.
Conspiracy theorists in Davos conference: "Saving the planet, The World is stuck on Stupid, "David Dilley, CEO of "Global Weather Oscillations."
Charles Schwab, at WEF in Davos: Wants to Master the Future.
Dan Bongino, Interviewed Bill Gates who owns four airplanes and bought "Offsets" from Al Gore, Bill Gates commented: My Jets are not part of the problem! February 11,2023
Dr, Fred Singer, "Hot Talk. Cold Science."
Dr Judith Curry: "Climate Uncertainty and Risk," "Thermodynamics of Atmospheres and Oceans."
Global Liberty Institute.com. "Climate Emergency Crisis."
Grant Ritchie: "Atmospheric Chemistry: From The Surface To The Stratosphere."
Jim Bohlen, Founder of Greenpeace International;" Climate Change based on fake narrative."

John Kerry: "The US Annual Climate Normals show: from 1901 until 2020 a rise of F 1 degree!"

Jack O' Malley, in Hot Talk: "The entire globalist movement is out to suppression of the people, Money is Power!"

Michael Schellenberger: "Apocalypse Never", Climate Zealotry," Why Environmental Alarmists Hurt us all."

Mark Morano's Video's: "The Climate Hustle."

National Geographic Almanac.

NASA's Goddard Institute for Space Studies (NASA/Giss)

NOAA, National Oceanic and Atmospheric Administration.

Oliver Heariside, Physicists. Explanation of Radio Waves.

Pete Hegseth, "Battle for the American Mind."

Professor Jerry McSerezze, Mathematician "Arctic Extend."

Paul R. Ehrlich, American Biologist on Climate Change: "any kind of weather, the world over reacts, but they need to get their act together and understand that it is all normal and totally cyclical."

Pavlova Holland, Geophysical Research.

Palmetto, Clean Energy Technology Platform, Solar Power.

Scmutz and P.L. Flint, "Erosions" 2009.

Sebastian Gorka, "The War for America's soul."

Steve Milloy, Overreach of Government on Gas stoves, causing Asthma and Cognitive Damage. "Gas use is attributed to Bogus Childrens Asthma."

UCAR, Center for Science Education.

A Personal Note from The Author.

As an experienced Horticulturist I have done a lot of research in order to write this book. With my personal conviction as a Christian, I know that God created this Earth for the benefit of the people. Because He gave us dominion over the Earth so we the people can take care of the Earth.

I am amazed how the Earth was created with such a perfection and detail so that we cannot change any basics of the Earth.

Many Scientists have studied how the Earth operates they have always tried to find reasons to explain how and why the Earth works as it does in all situations and with possible calamities, and it never really changes.

The facts remain because God is God and we are not.

The main reason behind Climate Change and Global Warming is that the Governments use fear tactics about it in order to have control over people and to stay in power.

Whether you are a believer or not, you can read all the facts from a human standpoint.

In the bible God mentioned the words: "Fear Not" 365 times which is once for every day. Also: "I will be with you always."

That is my motivation to write this book and you will be able to learn the truth about everything that happens on this Earth because it is all Cyclical and made with a perfect intention.

Signed by the Author, Dr. Dirk van Leenen

A book on a murder in the government of the European Union.
A gripping International Thriller.